# Exam Success
## Inspection and Testing 2394 and 2395
Malcolm Doughton and Chris Johnston

City & Guilds is the UK's leading provider of vocational qualifications, offering over 500 awards across a wide range of industries, and progressing from entry level to the highest levels of professional achievement. With over 8500 centres in 100 countries, City & Guilds is recognised by employers worldwide for providing qualifications that offer proof of the skills they need to get the job done.

Copies may be obtained from:
City & Guilds
1 Giltspur Street
London EC1A 9DD
For publications enquiries:
T +44 (0)844 543 0000
E-mail learningmaterials@cityandguilds.com

The Institution of Engineering and Technology is the new institution formed by the joining together of the IEE (The Institution of Electrical Engineers) and the IIE (The Institution of Incorporated Engineers). The new institution is the inheritor of the IEE brand and all its products and services including the IEE Wiring Regulations (BS 7671), the IEE Guidance Note 3 and supporting guidance material.

Copies may be obtained from:
The Institution of Engineering and Technology
PO Box 96
Stevenage
SG1 2SD, UK
T +44 (0)1438 767 328
E-mail sales@theiet.org
www.theiet.org

Candidates without this experience should consider undertaking the Level 2 Certificate in Fundamental Inspection, Testing and Initial Verification (2392-10) as well as obtaining industry experience. It is also strongly recommended that candidates should have achieved either the City & Guilds Certificate in the Requirements for Electrical Installations (BS 7671:2008) (2382-10) or a similar qualification demonstrating knowledge and understanding of the 17th Edition of BS 7671, Requirements for Electrical Installations, The IET Wiring Regulations.

The aim of this qualification is to enable the candidate to develop the necessary technical knowledge and understanding about the inspection, testing and certification of electrical installations. The candidate is assessed by one multiple choice examination and one written examination and this book is intended to provide help for these examinations. There is also a practical assessment.

### City & Guilds Level 3 Certificate in the Periodic Inspection, Testing and Certification of Electrical Installations (2395-01)

This qualification is intended for experienced people working in the electrical industry. Although City & Guilds does not state formal candidate entry requirements, the qualification is not intended for non-qualified electricians and/or those who do not have experience in inspecting, testing and reporting on the condition of electrical installations. Candidates without this experience should consider undertaking the Level 2 Certificate in Fundamental Inspection, Testing and Initial Verification (2392-10) as well as obtaining industry experience. It is also strongly recommended that candidates should have achieved either the City & Guilds Certificate in the Requirements for Electrical Installations (BS 7671:2008) (2382-10) or a similar qualification demonstrating knowledge and understanding of the 17th Edition of BS 7671, Requirements for Electrical Installations, The IET Wiring Regulations.

The aim of this qualification is to enable the candidate to develop the necessary technical knowledge and understanding about the inspection and testing of existing electrical installations and reporting on their condition. The candidate is assessed by one multiple choice examination and one written examination and this book is intended to provide help for these examinations. There is also a practical assessment.

### Finding a centre

In order to take the exam, you must register at an approved City & Guilds centre. You can find your nearest centre by looking up the qualification number 2394-01 or 2395-01 on www.cityandguilds.com. The IET (Institution of Engineering and Technology) is an accredited centre and runs exams in different parts of the country. For more details, see www.theiet.org.

**Notes**

At each centre, the Local Examinations Secretary will enter you for the award, collect your fees, arrange for your assessments to take place and correspond with City & Guilds on your behalf. The Local Examinations Secretary also receives your certificate and any correspondence from City & Guilds. Most centres will require you to attend a course of learning before entering you for the assessments. These are usually available as day or evening courses, over a number of weeks. Your local centre will advise you of their particular course availability.

### Awarding of certificates

When you undertake the City & Guilds 2394-01 or 2395-01 assessments, a certificate will be issued only when you have been successful in both the written and practical assessments. This certificate will not indicate a grade or percentage pass. The practical component also needs to be successfully completed before a certificate is issued.

Any correspondence is conducted through your centre. The centre will also receive a consolidated results list detailing the performance of all candidates entered for the written assessment at their centre. City & Guilds also provides notification of successful completion of the practical assessment to the centre. This is following the centre's submission of all candidates' successful completion of the practical assessments to City & Guilds.

If you have particular requirements that will affect your ability to attend and take the examinations, then your centre should refer to City & Guilds policy document 'Access to Assessment: Candidates with Particular Requirements'.

# The exam

The exam

# The exam in brief

The aim of these qualifications is to enable the candidate to develop the necessary technical knowledge and understanding about the inspection and testing of new work (2394-01) and periodic inspection of existing electrical installations (2395-01) and the completion of appropriate documentation.

For each qualification, candidates are required to complete the following assessments:
- a multiple choice assessment
- a written assessment
- a practical assessment.

### Unit 301 – Multiple choice examination

This examination covers areas of knowledge and understanding which are common to both initial verification and periodic inspection. This assessment is the same assessment for both qualifications and learners need only achieve the test once.

The examination comprises 40 questions and is 1 hour 20 minutes long. It is a 'closed book' examination, which means that you are not allowed to take any notes or reference books into the exam with you.

| | Topic/outcome | No of questions | % |
|---|---|---|---|
| 1 | Understand the requirements for completing the safe isolation of electrical circuits and installations | 7 | 18 |
| 2 | Understand the requirements for inspecting and testing electrical installations | 2 | 5 |
| 3 | Understand the requirements for safe testing of electrical installations | 11 | 27 |
| 4 | Understand the requirements for testing circuits which are not energised | 7 | 18 |
| 5 | Understand the requirements for testing energised installations | 13 | 32 |
| | **Total** | **40** | **100** |

### Unit 302 – Written Examination

The examination papers for both qualifications (2394 and 2395) have the same format. They each have 6 structured questions and are 1 hour 30 minutes long. Each question is worth 15 marks.

**Unit 303 – Practical assessment**

Each qualification has an associated practical assessment element that must be completed to obtain the 2394 and 2395 certificates. Full details can be obtained from your centre.

# Guidance on sitting the e-volve multiple choice examination

This section provides some useful information about the e-volve multiple choice examination for both qualifications (Unit 301). It considers:

- the format of the exam
- the structure of the questions
- effective methods of answering questions.

# Format of the exam

The examination is presented online and is arranged so that each question and the choice of answers are viewed separately, so only one question is viewed on the screen at any time.

# The structure of the questions

The questions are presented as a question stem, which may provide information and pose the actual question together with a choice of four possible answers, only one of which is correct.

# Effective methods of answering questions

In order to have the best chance of success it is essential to read the question carefully and make sure you understand what is being asked. The answers provided will include one correct answer and understanding the question is key to selecting the right response.

Based on their experience candidates sometimes believe that the answer they want to give is not included in the choices. In such instances the best approach is to consider the options given and select the one which is most appropriate for the question being asked.

One of the key features of the system is the ability to flag a question where you are unsure of the answer. This is registered by the system and a flag indicator is placed next to the question. This allows you to identify and return to flagged questions at any time during the exam.

**Notes**

One helpful approach is to work through the exam answering those questions to which the answers are readily apparent and flagging those which require more thought. It may be a good idea to guess these answers at this stage just in case you run out of time and are unable to return to all the questions you have flagged. When reaching the end of the questions you can then return to your flagged questions and use the remaining time to answer these questions. This helps to take some of the time related pressure away as the remaining time can be shared across just the flagged questions. Spending a long time over one question early in the examination puts the candidate under time pressure for the rest of the examination, so try to avoid this by restricting the amount of time spent on one question. As a guide, spend no more than four minutes on any one question, but remember you need to average two minutes per question overall, so you can only spend this additional time on a few questions.

Do not leave any questions unanswered as these will not gain any marks. Where the answer is not apparent and further thought does not help the next option is to eliminate those answers which cannot be correct. This will help reduce the number of right answer options. One way to do this is to look at each answer in turn. Ask yourself "is this possibly right or is it definitely wrong?" You will need to use the information contained in the question to help with this. The wrong answers you can then dismiss. Now have another look at the answers that you feel are possibly right and make your final decision.

## General guidance on sitting the written examinations 2394 and 2395

This section provides some useful information about the written examination for both qualifications (Unit 302) that you may find helpful. It considers:

- the format of the exam
- the structure of the exam
- how to interpret the questions and understand what is expected
- effective methods of answering questions.

It also identifies areas of the exam that are often not answered well and some of the most common errors that candidates make.

## Format of the exam

The layout of each paper consists of a question, or part question, followed by a space for the candidate to enter their answer. It is important that the answer is entered within the space provided and within the margins. The space provided for the answer is generous so that you have sufficient space to change your answer if you so wish. It is not an indication of how much you need to write. There is a total of ninety marks available for each of these papers. You are allowed 1 hour and 30 minutes to complete the examination and you are expected to answer all the questions. These examinations are "closed book", which means that you are not allowed to take any notes or reference books into the exam with you.

## Structure of the exam

Each paper is divided into two sections, Section A and Section B.

Section A of each paper has three questions which are often divided into a number of parts (a, b, c, and i), ii), iii) and so on), with each part of the question relating to a different learning outcome. The number of marks available for each part of the question is shown on the paper and this can be used to indicate how long to spend on your answer.

Section B also has three questions, but these relate to the scenario contained within a "source" document. These questions may also be divided into parts but often relate to a single learning outcome. Candidates are expected to display an in-depth knowledge of the particular subject. Typical examples include describing a test procedure or evaluating test results.

## Interpreting questions and understanding what is expected in the written examination

There are two key points that need to be considered when reading the question:

### Consider the number of marks available for the question or part question
This provides a valuable indication of the depth of the answer required. For example, a question which carries one mark will require a much simpler answer than one for which fifteen marks are available. Also the space provided for your answer gives an indication as to the amount you are expected to write or draw, but remember that the space is often generous.

### Read carefully and answer what the question actually asks
Often a question is answered incorrectly because of a failure to understand what is being asked and what is required. It is an easy trap to fall into under

**Notes**

exam conditions where you are under pressure. The danger here is that you may answer a question to which you know the answer, but which is not the question you have been asked in the paper. Remember, the questions are set to establish your level of understanding in specific areas, so the correct response is important if the marks are to be obtained. Take a little longer to read the question carefully to ensure you are quite clear about what is required.

## Wording of questions

The wording of a question, coupled with the number of marks available, gives valuable clues as to what is expected. The words used in the question provide the first clue. If you look out for the following words and phrases and understand what they mean, you should be able to provide an appropriate answer.

**State:** This means the answer is expected to be a short statement, not a long or rambling paragraph. The response to this type of question may even be just a single word or group of words which may not need to be a complete sentence.

**List:** This means you should produce a simple list of items or actions. The answer should be similar to that produced for the 'state' question. However, on this occasion the items would be expected to follow a sequence and form a list, as would be expected for a shopping trip.

**Explain briefly:** This requires a brief explanation; usually no more than one or two sentences. It does not require paragraphs of explanation and the word 'briefly' is used to indicate this requirement.

**Explain with the aid of a diagram:** This means exactly what is says. The answer should comprise both a diagram and an explanation. The examiner is trying to help you achieve maximum marks by asking for both an explanation and a diagram because this method of providing information is likely to be the most efficient.

**Show all calculations:** Again the examiner is trying to help you score as many marks as possible. Where a calculation is required and the only thing offered by the candidate is the numerical answer, then if it is wrong, the candidate would score no marks. If the candidate includes each step of the calculation then marks will be awarded for each correct step. It is always in your best interest to show all stages of the process. Where relevant remember to show the applicable units which apply to your answer e.g. V, Ω, kA etc.

**With the aid of a fully labelled diagram:** This indicates that a diagram needs to be provided with the component parts clearly labelled. The marks for these questions are divided between the diagram and the labelling.

To obtain the maximum marks for the question both the labelling and the diagram must be completed.

**Describe:** These questions often relate to test procedures and you are required to demonstrate your knowledge of the test process. Look at the number of marks available to give you an indication of how much detail you need to go into.

**Describe, in detail:** This indicates that a more detailed answer is required and again the number of marks available for the question gives an indication of the depth of the answer required.

A series of short bullet pointed statements is a very effective method of providing an answer, but remember that all necessary information must be included.

**Direct measurement:** This indicates that a test is required and the results are not to be established by using a calculation. For example, where you are asked to describe the direct measurement of earth fault loop impedance, then a description of the test procedure is required. Describing an $R_1 + R_2$ test and then stating how to determine the value by calculation using $Z_s = Z_e + (R_1 + R_2)$ will result in no marks being awarded for the answer.

## Terminology

It is important to answer questions using the correct terminology, which is the same as that used in Guidance Note 3 and BS 7671. Always use correct titles and terminology. Brand names should **not** be used to describe items of equipment, test instruments and the like. For example, the instrument used to test for continuity is a low resistance ohmmeter. It should not be referred to using a manufacturer's name (eg Megger) or referred to as a continuity tester. This is because the precise performance requirements given in BS 7671 and IET Guidance Note 3 must be met. Some continuity testers and instruments that provide continuity features may not meet these requirements.

Multifunction test instruments are commonly used for testing electrical installations. You need to be aware of the individual functions and ranges of these instruments. This includes such functions as insulation resistance, continuity, earth fault loop impedance and prospective fault current measurement. The appropriate measurement scale for a particular test must be clearly stated. Also be careful to use the correct units and symbols to describe test instrument readings (m$\Omega$, $\Omega$, M$\Omega$, A, kA, ms etc).

The use of the correct terminology for the component parts of an electrical system is also important. The application of the terms used in BS 7671 is necessary as this leaves no doubt as to the part being described. Typical terms include earthing conductor, main protective bonding conductors and circuit protective conductors. Terms such as 'earth wires' or 'cross bonding' do not correctly identify components and the examiner is unable to award marks for such items.

Another common error is the use of incorrect titles for documents. If you refer to the 'Electricity at Work Act' instead of the 'Electricity at Work Regulations' or the 'Health and Safety at Work Regulations' instead of the 'Health and Safety at Work Act', this will result in no marks being awarded. This also applies to the title of documents which are completed during the inspection and testing process and handed to the client.

## Common problem areas associated with Section B

Section B of each paper is related to a scenario which is given in the Source Document. The scenario gives details of an electrical installation, or part of an installation. It identifies what is to be carried out, and provides information that is to be used to answer the final three questions on the exam paper. This means that the answers to these questions should relate to the installation identified in the scenario.

The most common error made by candidates at this point in the paper is failing to read the scenario and apply the information given to these final three questions. For example, when the scenario clearly states a TN-C-S system is used and a question asks for a diagram of the earth fault loop path, producing a drawing of a TN-S system will result in no marks being awarded for that question.

It may be beneficial for you to highlight key pieces of information in the scenario. This will help you to concentrate on what you are reading and will make referencing information easier when you are answering the questions. This technique may also be useful when reading individual questions.

You are expected to be able to describe the procedures for carrying out activities, including the inspection and testing of installations and circuits. These descriptions should follow the format given in IET Guidance Note 3.

Common errors when answering questions that relate to the scenario are:
- not obtaining permission for isolation or for testing to proceed.
- no isolation procedure mentioned when it is appropriate
- no instrument and lead check carried out
- incorrect procedures described, such as not being able to describe the three steps in IET Guidance Note 3 for ring final circuit continuity
- not describing the test process in the correct sequence
- failure to consider the safety aspects necessary for the testing process
- failure to reinstate the installation safely once testing is complete.

## Confirmation of compliance

Some questions are included to establish your ability to confirm that measured test results meet the requirements of BS 7671. In order to do this you will be expected to show what steps are taken for this process and any calculations that may be involved. You will also be expected to identify appropriate action for any situations where the results do not meet the requirements. This will be different for initial verification (2394) compared with periodic inspection (2395).

A common area for error is the application of the 'rule of thumb' to the maximum tabulated values of earth fault loop impedance (these maximum values will be given in the scenario or question information) in order to compare these with the measured values. When candidates do not correctly apply the rule of thumb, it shows they are unable to correctly identify compliance with BS 7671 and therefore no marks are awarded.

## Schedules of test results

The section B questions and/or the source document may include details of test values obtained or a Schedule of Test Results may be provided. Questions may then relate to confirming the compliance of the installation or circuits based upon the information given on these documents and other given details such as maximum tabulated $Z_s$ values from BS 7671.

# Putting your answers on paper

There are many ways in which questions can be answered and you will need to find the method that best suits you. Here are some hints to help you decide on a style.

It is important to remember when you are answering the questions that the examiners cannot:

- ask you further questions to establish your understanding – they can only award marks for the information you provide
- assume what you mean or know – they can only interpret the information they are given in your answers.

Due to the time constraints of the exam, do **not** waste time by copying out the question. The question you are answering is directly above the space where you are writing your answer, so the examiner can see the question when marking, just as you can see it while completing your answer.

The marking of your answers does not include any penalties or additional marks for spelling or grammar. However, the examiners will still need to be able to read your answers and your handwriting, so write as clearly as possible.

The completed scripts are scanned so they can be read and marked electronically and blue or black ink provides the clearest text and pencil should not be used for text. Providing the examiners can understand your response (and the answer is correct), then marks will be awarded.

The answers do not need to be in the form of an essay or long descriptive text. A simple bullet pointed, step-by-step approach to the answer provides you with an easy reference as to what you have included when you read back through the answer. It also allows you to insert any item you have missed and you can indicate the correct location to the examiner.

# Outcome 6 – Understand the requirements for testing before circuits are energised

You are required to be able to:

**6.1**  State why it is necessary to verify the continuity, to include:
- protective bonding conductors
- circuit protective conductors
- ring final circuit conductors

**6.2**  State the methods for verifying the continuity, to include:
- protective conductors
- ring final circuit conductors

**6.3**  Explain the factors that effect conductor resistance values.

**6.4**  Specify the procedures for completing insulation resistance testing

**6.5**  State the effects on insulation resistance values that the following can have:
- cables connected in parallel
- variations in cable length

**6.6**  Explain why it is necessary to verify polarity

**6.7**  State the procedures for verifying polarity

# Outcome 7 – Understand the requirements for testing energised installations

You are required to be able to:

**7.1**  State the methods of measuring earth electrode resistance to include:
- installations forming part of a TT system
- generators and transformers

**7.2**  Describe common earth fault loop paths

**7.3**  State the methods for verifying protection by automatic disconnection of supply

**7.4**  Identify the requirements for the measurement of prospective fault current

**7.5**  Specify the methods for determining prospective fault current

**7.6**  Verify the suitability of protective devices for prospective faults currents

**7.7**  Specify the methods for testing the operation of residual current devices

**7.8** State the reasons for verifying phase sequence
**7.9** State the methods used to verify phase sequence
**7.10** Describe the methods used to verify voltage drop
**7.11** State the cause of voltage drop in an electrical installation
**7.12** Determine voltage drop
**7.13** State the need for functional testing
**7.14** Identify items which require functional testing
**7.15** State the appropriate procedures for dealing with clients during the periodic inspection process

## Outcome 8 – Understand and interpret test results

You are required to be able to:
**8.1** Explain why it is necessary to confirm whether test results comply with standard values
**8.2** Analyse test results to determine action to be taken

## Outcome 9 – Understand the requirements for the completion of electrical installation condition reports and associated documentation

You are required to be able to:
**9.1** Explain the purpose of an electrical installation condition report
**9.2** State the information that must be contained within an electrical installation condition report
**9.3** Explain the requirements for the recording and retention of completed electrical installation condition reports, in accordance with BS 7671
**9.4** Identify appropriate methods for providing information to a client following completion of the electrical installation condition report

## Outcome 10 – Be able to confirm safety of system and equipment prior to completion of inspection and testing

You are required to be able to:
**10.1** Carry out safe isolation procedures in accordance with regulatory requirements

| | |
|---|---|
| **10.2** Comply with the health and safety requirements of themselves and others within the work location during the periodic process | **Notes** |
| **10.3** Check the safety of electrical systems prior to the commencement of inspection and testing | |

## Outcome 11 – Be able to carry out inspection of electrical installations

You are required to be able to:

**11.1** Identify a safe system of work appropriate to the work activity

**11.2** Carry out a periodic inspection of an electrical installation in accordance with the requirements of BS 7671 and IET Guidance Note 3

**11.3** Complete a Condition Report Inspection Schedule in accordance with BS 7671 and IET Guidance Note 3

## Outcome 12 – Be able to test electrical installations in service

You are required to be able to:

**12.1** Select the test instruments and their accessories for tests to include:
- continuity
- insulation resistance
- polarity
- earth electrode resistance
- earth fault loop impedance
- prospective fault current
- RCD operation
- phase sequence
- functional testing

**12.2** Carry out tests in accordance with BS 7671 and Guidance notes 3 to include:
- continuity including
  - main protective bonding conductors
  - circuit protective conductors
  - Ring Final Circuits
- insulation resistance
- polarity
- external earth fault loop impedance ($Z_e$)
- system earth fault loop impedance ($Z_s$)
- prospective fault current
- RCD operation including additional protection

- phase sequence
- functional testing
- additional protection
- verification of voltage drop

**12.3** Compare test results with standard requirements and previous test results

## Outcome 13 – Produce a condition report with recorded observations and classifications

You are required to be able to:

**13.1** Use information to determine defects and non-compliances to include:
- dwellings
- other premises

**13.2** Complete Electrical Installation Condition Report and associated documents

**13.3** Handover of the condition report to the client with appropriate information and guidance regarding actions to be taken

# E-volve multiple choice examination (2394-301 and 2395-301)

E-volve multiple
choice exam

# Sample test 2394/5-301

The sample test below has 40 questions, the same number as the online exam, and its structure follows that of the online exam. The test appears first without answers, so you can use it as a mock exam. It is then repeated with answers and comments from our examiners.

Answer the questions by filling in the circle next to your chosen option.

**1  In order to undertake the inspection and testing of an electrical installation a person must be a:**

- ○ a    competent person
- ○ b    principle person
- ○ c    duty person
- ○ d    registered person.

**2  The term 'Dutyholder' used in the Electricity At Work Regulations refers to the person:**

- ○ a    making the final payment on completion of the work
- ○ b    responsible for the electrical installation
- ○ c    reviewing the documents
- ○ d    ordering the work.

**3  Which of the following identifies the correct process for safe isolation of a circuit protected by a circuit breaker?**

- ○ a    Switch off, post notices.
- ○ b    Switch off, lock off, post notices.
- ○ c    Switch off, confirm isolation, post notices.
- ○ d    Switch off, lock off, confirm isolation, post notices.

**4  Work is to be carried out on an installation which includes a Solar PV supply system. Which of the following <u>must</u> be carried out before isolating the DNO's supply to the installation?**

- ○ a    Isolate the Solar PV supply.
- ○ b    Disconnect the installation loads.
- ○ c    Close all the switches.
- ○ d    Disconnect the bonding conductors.

39 Following a test of earth fault loop impedance ($Z_s$) the results are compared with the values given in BS 7671. Which of the following describes the purpose of this comparison?

- a   To confirm correct test methods are used.
- b   To confirm conductor csa is suitable for the circuit.
- c   To determine whether disconnection times will be achieved under earth fault conditions.
- d   To determine whether the protective device will operate under short circuit conditions.

40 Checking that the measured test results meet the required values will enable the inspector to confirm that the electrical installation is

- a   correctly installed
- b   safe to be in service
- c   compliant with BS 5266
- d   never going to be dangerous.

**Notes**

# Questions and answers

The questions in the sample test are repeated below with answers and comments from our examiners.

**1  In order to undertake the inspection and testing of an electrical installation a person must be a:**

◉ a    competent person
○ b    principle person
○ c    duty person
○ d    registered person.

### Comments
Regulation 16 of the Electricity at Work Regulations requires persons to be competent to prevent danger. As such danger can occur when carrying out an inspection and test on an electrical installation, the inspector is required to be competent. This is further reinforced in GN 3 as certification and reports are required to be signed by a competent person.

**2  The term 'Dutyholder' used in the Electricity At Work Regulations refers to the person:**

○ a    making the final payment on completion of the work
◉ b    responsible for the electrical installation
○ c    reviewing the documents
○ d    ordering the work.

### Comments
The Electricity at Work Regulations identifies anyone taking control of, or responsibility for, an electrical system as a Dutyholder (Memorandum of Guidance Regulation 3 Item 55). Dutyholders have a duty to ensure the safety of themselves and others.

**3  Which of the following identifies the correct process for safe isolation of a circuit protected by a circuit breaker?**

○ a    Switch off, post notices.
○ b    Switch off, lock off, post notices.
○ c    Switch off, confirm isolation, post notices.
◉ d    Switch off, lock off, confirm isolation, post notices.

### Comments
All of the steps listed in option d are required to complete isolation and to ensure the circuit is safe to work on.

**4** Work is to be carried out on an installation which includes a Solar PV supply system. Which of the following __must__ be carried out before isolating the DNO's supply to the installation?

- ⦿ a    Isolate the Solar PV supply.
- ○ b    Disconnect the installation loads.
- ○ c    Close all the switches.
- ○ d    Disconnect the bonding conductors.

**Comments**

Solar PV systems generate even at low light levels and to ensure the safety of the inspector and those persons in the vicinity the system must be isolated. Solar PV systems are an alternative source of supply. Although many Solar PV systems will shut down automatically following loss of supply, this should never be taken for granted and isolation of the PV system should always be carried out.

**5** A test of earth fault loop impedance is to be carried out in an area which is accessible to the public. Which of the following is __not__ a suitable method of protecting the public from the risk of electric shock?

- ○ a    Barriers.
- ○ b    Warning signs.
- ⦿ c    Safe isolation.
- ○ d    Restricting access.

**Comments**

Safe isolation is not appropriate in this instance as the test is a live test and so has to be carried out with the supply connected.

**6**  Work is to be carried out at the distribution board supplying a multi-storey office building. Which of the following must be placed out of service, to ensure the safety of the users, before the safe isolation of the distribution board is carried out?

- ○ a    General lighting circuits.
- ○ b    Air conditioning system.
- ⦿ c    Lifts to all floors.
- ○ d    Burglar alarms.

**Comments**

The client's agreement to isolate the lift supplies must be obtained and the lifts taken out of service to prevent persons being trapped in the lifts when the supply to the distribution board is isolated.

7  Who is <u>most</u> at risk when an inspector fails to isolate a lighting circuit before removing a luminaire cover to carry out an inspection?

&#9673; a    The inspector.
&#9711; b    Other operatives.
&#9711; c    The general public.
&#9711; d    The client.

**Comments**
As the work is being carried out by the inspector and the other persons listed are unlikely to have access to the luminaire the inspector is at most risk when isolation is not carried out prior to dismantling.

8  Which publication specifically identifies the need for electrical installations to be safe for use and maintained in that condition?

&#9711; a    HSE Guidance GS 38.
&#9673; b    Electricity at Work Regulations.
&#9711; c    The Health and Safety at Work (etc.) Act.
&#9711; d    Electricity Safety, Quality and Continuity Regulations.

**Comments**
Regulation 4 of the Electricity at Work Regulations requires electrical systems to be installed and maintained so far as is reasonably practical to prevent danger.

9  Which publication specifically identifies the requirements for test leads and probes to be used when carrying out tests at 230 V ac?

&#9711; a    BS 7671.
&#9673; b    HSE Guidance GS 38.
&#9711; c    Electricity at Work Regulations.
&#9711; d    Provision and Use of Work Equipment Regulations.

**Comments**
HSE GS 38 provides guidance on the test equipment used for circuits with rated voltages not exceeding 650 V.

16 Test instruments should be calibrated in accordance with the manufacturer's instructions to ensure accurate test results. In addition to calibration which of the following should be carried out and recorded?

○ a   Regular confirmation of the instrument's compliance with BS 7671.
○ b   Extension of the instrument warranty period.
◉ c   Regular instrument accuracy checks.
○ d   Closed circuit voltage checks.

### Comments
Regular instrument accuracy checks are required to ensure the instrument continues to provide accurate results, as identified in GN 3 4.2. If regular checks are not carried out it may result in the need to re-inspect and test earlier installations, once an instrument defect is identified.

17 Which of the following is the maximum voltage above which test leads must comply with the requirements of GS 38?

○ a   25 V a.c.
○ b   25 V d.c.
◉ c   50 V a.c.
○ d   50 V d.c.

### Comments
Where the test voltages are $\leq 50$ V a.c. the requirement for GS 38 compliant leads is not necessary where the short circuit fault current is unlikely to cause a high energy flashover. Test current for this test is $\geq 200$ mA.

18 Which of the following requirements from GS 38 applies to test equipment which requires more than one test lead?

○ a   The leads must be at least 1 m in length.
○ b   The leads must have a csa of at least 1.5 mm².
◉ c   The leads must be colour coded for identification.
○ d   The leads must be permanently attached to the instrument.

### Comments
Item c is the only one of the listed items described in GS 38.

**Notes**

**19 Which of the following correctly describes the purpose of a continuity of ring final circuit conductors test?**

- ◉ a   To confirm the conductors are installed as a complete ring.
- ○ b   To confirm polarity is correct at each point on the circuit.
- ○ c   To determine the $R_1 + R_2$ value for the final circuit.
- ○ d   To confirm the conductors are the correct csa.

**Comments**

Of the four options given, a is the only one which identifies the **purpose** of the test as described in BS 7671 (Regulation 612.2.2). The other three may be determined as a result of the test but are not the purpose of carrying out the test.

**20 A test for continuity of circuit protective conductors is carried out at every point on a circuit. Which of the following describes the purpose of this test?**

- ○ a   To confirm correct polarity.
- ○ b   To confirm the conductors are the correct csa.
- ○ c   To determine the $R_1 + R_2$ value for the circuit.
- ◉ d   To confirm exposed conductive parts are connected to the MET.

**Comments**

Of the four options given, d is the only one which identifies the **purpose** of the test. The other three may be determined as a result of the test, depending on the method used, but are not the purpose of carrying out the test.

**21 Which of the following factors directly affects the conductor resistance of a cable?**

- ○ a   Insulation and csa.
- ◉ b   Length and csa.
- ○ c   Length and insulation.
- ○ d   Load current and csa.

**Comments**

Of the four options given, b is the only one in which both components will directly affect the conductor resistance. Load current will only have an indirect effect on conductor resistance by raising the conductor temperature.

Notes

31 Which of the following is <u>not</u> a method for determining prospective earth fault current at the origin of an installation?

○ a   Enquiry to the DNO.
○ b   Direct measurement.
○ c   Calculation using measured $Z_e$ and the supply voltage.
◉ d   Calculation using the measured $Z_s$ and the supply voltage.

**Comments**
Prospective fault current is measured to determine the maximum fault current which may occur. Any impedance measurement taken at the end of a circuit will be higher than that at the origin of the circuit. As a result the value obtained from the calculation in d is not going to produce the maximum value.

32 Which of the following correctly describes the relationship between prospective fault current and the fault current rating of a BS EN 60898 circuit breaker?

○ a   $I_{pf} \geq I_{cn}$.
◉ b   $I_{pf} \leq I_{cn}$.
○ c   $I_{pf} \geq I_n$.
○ d   $I_{pf} \leq I_n$.

**Comments**
The $I_{cn}$ of the circuit breaker is the maximum current the circuit breaker can safely disconnect, although it may then no longer be serviceable. To meet the requirements this value must be equal to or higher than the maximum fault current which may occur.

33 Which of the following protective devices is <u>never</u> suitable for use to disconnect fault currents in excess of 10 kA?

◉ a   BS 3036 fuse.
○ b   BS 88-3 fuse.
○ c   BS 88-2 fuse.
○ d   BS EN 60898 circuit breaker.

**Comments**
The maximum breaking capacity of a BS 3036 fuse is 4 kA, so it is not a suitable protective device where the fault current is 10 kA.

**34** A 100 mA BS EN 61008 RCD is installed in a TT installation to provide fault protection. Which of the following identifies the maximum test current to be applied and the maximum disconnection time at that test current when testing the RCD?

- ○ a   100 mA and 200 ms.
- ⦿ b   100 mA and 300 ms.
- ○ c   500 mA and 200ms.
- ○ d   500 mA and 300 ms.

**Comments**

The RCD is provided for fault protection and not additional protection and the 1 x $I_{\Delta n}$ is the maximum test current that needs to be applied. As this is a BS EN 61008 device the maximum disconnection time is 300 ms. A maximum disconnection time of 200 ms applies to some older RCDs manufactured to a British Standard, but not to BS EN 61008 devices.

**35** A 30 mA BS EN 61009 RCBO is installed on a socket-outlet circuit to provide additional protection. Which of the following identifies the maximum test current to be applied and the maximum disconnection time at that test current?

- ○ a   30 mA and 300 ms.
- ○ b   30 mA and 200 ms.
- ○ c   150 mA and 300 ms.
- ⦿ d   150 mA and 40 ms.

**Comments**

The RCBO is provided for additional protection and the 5 x $I_{\Delta n}$ is the maximum fault current that needs to be applied. As this device is providing additional protection, the maximum disconnection time is 40 ms, as required by BS 7671 Regulation 414.1.1

**36** Which of the following is <u>not</u> a suitable method of confirming phase sequence?

- ○ a   Measurement using a rotating disc type instrument.
- ○ b   Measurement using an indicator type instrument.
- ⦿ c   Measurement using an approved voltage indicator.
- ○ d   Checking polarity and connections throughout the installation.

**Comments**

The methods of determining phase sequence are given in GN 3. An approved voltage indicator cannot be used to determine phase sequence.

37 **The purpose of a functional test on an RCD carried out quarterly, using the integral test button, is to confirm:**

○ a   The correct functioning of the RCD in the event of a fault
○ b   That the operation of the RCD complies with BS 7671
○ c   That the RCD operates in the event of a short circuit
◉ d   The correct mechanical operation of the RCD mechanism.

**Comments**
Operating the RCD test button will only confirm the mechanical operation of the RCD.

38 **A test is carried out on the interlock for switching an alternative power supply. Which of the following is the type of test being undertaken?**

◉ a   Functional.
○ b   Continuity.
○ c   Load.
○ d   Installation.

**Comments**
This test confirms the operation of the interlocking device and is therefore a functional test.

39 **Following a test of earth fault loop impedance ($Z_s$) the results are compared with the values given in BS 7671. Which of the following describes the purpose of this comparison?**

○ a   To confirm correct test methods are used.
○ b   To confirm conductor csa is suitable for the circuit.
◉ c   To determine whether disconnection times will be achieved under earth fault conditions.
○ d   To determine whether the protective device will operate under short circuit conditions.

**Comments**
The operation of a protective device under earth fault conditions relies upon a sufficiently large current flowing. From Ohm's Law the supply voltage and impedance of the system will determine the fault current. The $Z_s$ value is the variable and can therefore be used to determine whether disconnection will be achieved within the required time.

Notes

**40 Checking that the measured test results meet the required values will enable the inspector to confirm that the electrical installation is:**

○ a    correctly installed

◉ b    safe to be in service

○ c    compliant with BS 5266

○ d    never going to be dangerous.

**Comments**

Providing the test results obtained during the inspection and testing meet the requirements of BS 7671 and the design requirements, the installation will be suitable for use.

**WRITTEN EXAMINATION 2394-302**
**PRINCIPLES, PRACTICES AND LEGISLATION FOR THE INITIAL
VERIFICATION OF ELECTRICAL INSTALLATIONS (2394-302)**

67

This answer and similar variations are perfectly acceptable and you would obtain the marks for the answer. However, it does rely on the procedure being written down correctly the first time without missing any steps. It involves a considerable amount of writing and it is often difficult to spot any errors during the examination when you read back through your answer.

## Option 2

The answer can be structured as a set of bullet points, as follows.

- *Once the socket-outlets are terminated prepare the cable for termination at the distribution board*
- *Link line and cpc at the db*
- *Select a low resistance ohmmeter*
- *Check that calibration is current, leads are in good condition, batteries are OK and the instrument is functioning*
- *Zero the test leads (or Null the test leads)*
- *Using a suitable plug adaptor, test between line and cpc at each socket-outlet*
- *Record the highest value. This would be the reading at the furthest socket-outlet from the db*

There is plenty of space to write down your answer, so you can leave a space between each bullet point, giving you the opportunity to add an additional line in later if you have missed a step in the process. You can also read through your answer much quicker and identify any omissions.

Additional bullet points may be added at the end if you find you have missed out some information providing they identify when the action is carried out. For example if you find you have omitted to null the test leads the final bullet point could be added as:

- *Before carrying out the test the test leads should be nulled*

This would then ensure that the appropriate marks are awarded for your answer.

You will notice that there are some abbreviations used, such as 'db', and 'cpc'. The examiner will understand these when used in this context and it will not affect the marks awarded for the answer given.

The second option, with the use of the bullet points, is generally quicker to complete, less prone to errors or omissions and clearly demonstrates your understanding of the test process. You can try both methods to find out which best suits you. However, Option 2 is strongly recommended.

**Notes**

It is important to recognise the need to carry out testing in a way that does not put yourself or anyone else at risk. Also the test must be carried out in such a manner that produces valid results. No marks will be awarded for an answer that describes dangerous techniques and a heavy penalty will be applied to answers that describe invalid procedures.

The installations being inspected and tested are generally not energised as this qualification is about inspecting and testing new work. Where an addition or alteration is to be carried out on an existing installation there will be a need to isolate the installation before certain work is carried out. Therefore a question on the paper could ask for safe isolation to be carried out.

# Sample test 2394-302

The sample exam paper below has six questions, the same number as the written examination, and follows the structure of the exam. The exam appears with spaces for you to write your answers so you can use it as a mock exam, if you run out of space continue on an additional sheet of paper. The exam is then repeated with worked-through answers and comments. The front cover of the examination paper is shown below, identifying the requirements of the exam. The date of the exam would normally appear on the front sheet and the heading "Exam Success" would be replaced with the examination series. The remaining information appears as it would on the examination paper.

||||| 2 3 9 4 3 0 2 0 6 1 3

**City & Guilds**

Exam Success 2394-302

**Principles, practices and legislation for the initial verification of electrical installations**
Initial inspection, testing and certification

| If provided, stick your candidate barcode label here. | Date of examination 18:30 – 20:00 |
|---|---|

Candidate name (first, last)
First

Last

Candidate enrolment number    Date of birth (DDMMYYYY)    Gender (M/F)

Assessment date (DDMMYYYY)    Centre number    Candidate signature and declaration*

- If any additional answer sheets are used, enter the additional number of pages in this box. ➡ ☐ ☐
- Please ensure that you **staple** additional answer sheets to the **back** of this answer booklet, clearly labelling them with your full name, enrolment number, centre number and qualification number in BLOCK CAPITALS.
- If provided with source documents, these documents **will not** be returned to City & Guilds, and will be shredded. **Do not** write on the source documents.
**\*I declare that I had no prior knowledge of the questions in this assessment and that I will not divulge to any person any information about the questions.**

**You should have the following for this assessment:**
- non-programmable calculator        • one enclosed source document
- a pen with black or blue ink
- drawing instruments

**General instructions**
- This examination consists of **six** structured questions. Candidates must answer **all six** questions:
- Section A – **three** structured questions, each carrying 15 marks.
- Section B – **three** structured, scenario-based, questions each carrying 15 marks.
- The terminology used in answering this paper should be in accordance with current IET Publications.
- Show **all** calculations. If you use a calculator, show sufficient steps to justify your answers.
- Write all your working out and answers in this booklet.

**Section A –** All questions carry equal marks. Answer **all three** questions. Show **all** calculations.

1  a)  i)  State **one** statutory document to which the inspector may refer whilst carrying out an inspection and test of an electrical installation.

_____

_____

1 mark

ii)  State **two** non-statutory publications to which the  inspector may refer whilst carrying out an inspection and test of an electrical installation.

_____

_____

_____

_____

2 marks

iii)  State the title given in law to the inspector whilst carrying out the inspection of an electrical installation.

_____

_____

1 mark

iv)  State the legal status of the inspector.

_____

_____

1 mark

b)  State the scope of the Minor Electrical Installation Works Certificate.

_____

_____

_____

_____

2 marks

c)  State what must be verified when inspecting electrical  equipment, during an initial verification.

_____

_____

_____

_____

_____

_____

3 marks

## Figure 2

| Conductor size mm² | Resistance in mΩ/m at 20 °C |
|---|---|
| 1.5 | 12.10 |
| 2.5 | 7.41 |
| 10 | 1.83 |

# Questions and answers

The questions in sample test 2394-302 are repeated below with sample answers, and comments and advice where appropriate.

## Section A

1 a) i) State **one** statutory document to which the inspector may refer whilst carrying out an inspection and test of an electrical installation. (1 mark)

**Answer**
The Electricity At Work Regulations 1989

**Comments**
Because the question has asked for **one** answer, only the first answer given by the candidate would be considered. This is because candidates sometimes write down everything they can remember in the hope that one of the answers may be correct. This is not showing knowledge or understanding.

"EaWR" or "EWR" are acceptable. Electricity at Work "Act" is incorrect.

The candidate could have offered other health and safety regulations or reference to the Health and Safety at Work Act as a possible answer. The particular document must relate to the process of inspection and must be a statutory document.

1 a) ii) State **two** non-statutory publications to which the inspector may refer whilst carrying out an inspection and test of an electrical installation. (2 marks)

**Answer**
BS 7671:2008(2011) and IET Guidance Note 3

**Comments**
Because the question has asked for **two** answers, only the first two answers given by the candidate would be considered. Other Guidance Notes would be acceptable but the most relevant is Guidance Note 3.

The On-Site Guide is a perfectly acceptable alternative answer.

1  a)  iii)  State the title given in law to the inspector whilst carrying out the inspection of an electrical installation.    (1 mark)

**Answer**
Dutyholder

**Comments**
This question refers to the requirements of The Electricity at Work Regulations and the term "Dutyholder" refers to anyone having a duty of care to others.

1  a)  iv)  State the legal status of the inspector.    (1 mark)

**Answer**
Competent

**Comments**
Regulation 16 of The Electricity at Work Regulations requires persons to be competent when carrying out any activity on an electrical system where the work may give rise to danger.

1  b)  State the scope of the Minor Electrical Installation Works Certificate.    (2 marks)

**Answer**
Alteration or addition to an existing circuit

**Comments**
A key part of this answer is the reference to "an existing circuit". An alteration or addition to an existing installation is NOT within the scope of a Minor Electrical Installation Certificate.

Any description covering the essential points "alteration" or "addition" and "existing circuit" would get the marks.

**Notes**

1  c)  State what must be verified when inspecting electrical equipment, during an initial verification.                    (3 marks)

**Answer**
Equipment is to British Standards
Equipment is correctly selected and erected
Equipment is not damaged

**Comments**
The question refers to the inspection of electrical equipment and NOT to the inspection of an electrical installation. BS 7671 refers specifically to what must be verified when inspecting electrical equipment.

"Equipment is installed to the requirements of BS 7671" is an acceptable alternative to "correctly selected and erected" as this is a requirement of The Wiring Regulations. It is NOT an acceptable alternative to "Equipment is to British Standards", but reference to European or International standards would also be acceptable.

1  d)  List **five** characteristics of the supply which must be recorded on an Electrical Installation Certificate.                    (5 marks)

**Answer**
- Earthing arrangements
- Number and type of live conductors
- Nominal voltage
- Nominal frequency
- PFC

**Comments**
Because the question has asked for **five** answers, only the first five answers given by the candidate would be considered.

Acceptable answers must be those characteristics that are recorded on an Electrical Installation Certificate under the heading "Supply characteristics and Earthing Arrangements".

" Type of earthing", "external earth fault loop impedance", "$Z_e$", "ac/dc",. "Nature of supply" would gain a mark but only if none of the other answers given by the candidate included items from that section.

"Protective device", "main fuse" and "earth loop impedance" would not score marks because they are vague.

2  a)  i)   List **five** items to be inspected on a PVC conduit system
             prior to the installation of cables.                    (5 marks)

Notes

**Answer**
- Conduit system overall
- Fixings
- Bends
- Draw-in points
- Joints

2  a)  ii)  Explain what is being inspected for **each** of the items
             identified in a) i) above.                              (5 marks)

**Answer**
- Conduit system overall - conduit system is complete
- Fixings - Sufficient saddles fixed correctly
- Bends - Bending radius not too tight and bend not damaged
- Draw-in points - Sufficient number so that cables can be drawn in
  without damage
- Joints - are secure and fit correctly between lengths and at boxes

**Comments**

This question is in two parts. i) asks for what is to be inspected and ii)
requires the candidate to identify precisely the nature of the inspection.
Sometimes these two parts are combined into one question.

It is important for the candidate to state clearly what is being inspected
and what in particular they would be looking for.

Because the question has asked for **five** answers, only the first five
answers given by the candidate would be considered. The answer given in
ii) should match to the answer given in i).

"Conduit system not damaged", "conduit complies with appropriate
British Standard", "expansion joints fitted where necessary" and "bushes
are tight" are all acceptable answers.

Any answer relating to cables, cable terminations and conduit capacity
would not be acceptable, although the capacity of the conduit could
have been addressed by reference to the conduit being the correct size
specified by the designer.

**Notes**

2 b) The changing rooms at a sports club contain showers.
   i) State the position of Zone 2 (1 mark)

**Answer**
Extending to 600 mm from edge of Zone 1

**Comments**
A precise description of the zone is not required as this is a "closed book" exam. The examiner is looking for some recognition by the candidate as to where the zone sits within the shower area.

2 b) ii) State the minimum IP code for electrical equipment located in Zone 1 (1 mark)

**Answer**
IPX4

**Comments**
Although any IP code above IPX4 would be suitable, the question clearly states "minimum" so this is the only correct answer.

2 b) iii) State the type of switch that may be installed in Zone 2. (1 mark)

**Answer**
SELV Switch

**Comments**
A 230 V cord-operated ceiling switch is not permitted in Zone 2. The cord may enter the zone, but the switch must be outside of the zone.

2 c) List **two** labels that would normally be present on a newly completed domestic installation. (2 marks)

**Answer**
- Next inspection notice
- Earthing and bonding notice

**Comments**
The key word in this question is "normally". The examiner is looking for labels that are "normally" present on a new domestic installation. Answers that include labels that, under certain conditions are present, but are not "normally" present would not be acceptable.

Again, because the question has asked for **two** answers, only the first two answers given by the candidate would be considered.

An RCD test notice would also be an acceptable answer.

3  a)  Explain, in detail, why an earth fault loop impedance test would need to be carried out on existing circuits after the changing of a consumer unit within a domestic installation.  (5 marks)

**Answer**

Existing circuit protective devices have been removed and new protective devices have been fitted. The characteristics of the new devices are likely to be different from previous devices. Therefore it must be confirmed that the earth fault loop impedance is low enough so that the required disconnection times will be met.

**Comments**

The question asks for an explanation "in detail" so the description needs to be comprehensive if full marks are to be awarded. Answers that referred to disconnection and re-connection of conductors and the reliability of those connections would score marks.

3  b)  i)  Explain why the earthing conductor in an installation must be disconnected from the MET when measuring $Z_e$.  (2 marks)

**Answer**

To remove parallel earth paths so that the intended fault path can be confirmed to be reliable.

**Comments**

If the parallel earth paths are not removed during the test then the reliability of the test result is in doubt.

3  b)  ii)  Explain why the earthing conductor is connected to the MET when measuring prospective earth fault current.  (2 marks).

**Answer**

This is the condition that exists at the time of a fault and therefore is the only way to determine the maximum fault current to earth when carrying out the test.

**Comments**

In this case the purpose of the test is to determine the maximum prospective earth fault current. This will occur when the installation is energised and all earthing arrangements are in place, so the test must be carried out under these conditions.

**Notes**

3  c)  i)  List the measurements to be taken, at the main switch
            of a three phase TN-S system, in order to determine
            the installation prospective fault current ($I_{pf}$).      (3 marks)

**Answer**
- L1 to L2
- L1 to L3
- L2 to L3

**Comments**
There is no need to carry out any other tests to determine PFC because on
a three-phase system the largest fault current will be due to a symmetrical
short circuit, that is a short circuit between all line conductors at the same
time. A test between line conductors is an acceptable approximation.  If
additional tests were included in the answer then this would not make the
answer invalid.

The measurement of L1 to N, L2 to N and L3 to N would be an acceptable
Alternative method (see answer to 3cii) below).

3  c)  ii)  Explain how the recorded value of prospective fault
            current ($I_{pf}$) is determined following the test in
            c) i) above.                                            (3 marks)

**Answer**
The highest value of the three measurements

**Comments**
3 c) i) and ii) are linked and the answer to ii) is dependent on the test
method used in i). If the answer to i) was to measure between L and N
then this answer would be "the largest value x 2". If the answer to i)
included tests to earth then this answer must state "the largest line to
neutral value x 2".

## Section B

Remember that the answers to the following questions must relate to the scenario contained in the Source Document.

4) a) The loop length for the office ring final circuit is 60 m and all the socket-outlets are connected directly into the ring. Determine, showing **all** calculations

i) the expected $R_1 + R_2$ test value                    (7 marks)

**Answer**

$$r_1 + r_2 = \frac{60 \times (7.41 + 12.10)}{1000} = 1.17 \, \Omega$$

$$R_1 + R_2 = \frac{1.17}{4} = 0.293 \, \Omega$$

**Comments**

There is more than one way to determine this value but any correct method would be given the marks. The use of a temperature correction factor in this calculation is not appropriate because the question asks for an expected "test" value and the Source Document states that testing is to be carried out at 20 °C, which is the same temperature as that which applies to the mΩ/m values in Figure 2.

The calculation could have also been laid out as shown below

$$r_1 = \frac{60 \times 7.41}{1000} = 0.445 \, \Omega$$

$$r_2 = \frac{60 \times 12.10}{1000} = 0.726 \, \Omega$$

$$R_1 + R_2 = \frac{0.445 + 0.726}{4}$$

$$= 0.293 \, \Omega$$

**Notes**

4  a)  ii)  the expected measured $Z_s$ value                  (4 marks)

**Answer**

$Z_s = Z_e + R_1 + R_2$

$\phantom{Z_s} = 0.11 + 0.29$

$\phantom{Z_s} = 0.40\ \Omega$

**Comments**

The answer to this question is dependent upon the answer given in a) i) above. It is not normal practice to penalise a candidate twice for the same error. With this in mind, marks would be awarded for the correct formula, the correct value for $Z_e$, the value determined when answering a) i) and the correct total, including units.

4  b)  When measuring $Z_s$ for the toilet lighting circuit the circuit breaker operates. State why the circuit breaker tripped during the earth fault loop impedance test.                  (4 marks)

**Answer**

The test current may be greater than the instantaneous tripping current of the circuit breaker causing it to operate

**Comments**

The examiner is looking for any statement that indicates an understanding of what has happened. Reference to an RCD tripping would not gain any marks as the circuit is not protected by an RCD. This information is shown in Figure 1 in the Source Document.

5  The circuit for the three phase saw bench is installed to the local isolator for the saw.

a)  A test of earth fault loop impedance is to be carried out on the saw circuit.

i)  State the test instrument to be used.                  (2 marks)

**Answer**

Use an earth fault loop impedance tester

5  a)  ii)  State the document that specifies the requirements for the test leads                  (2 marks)

**Answer**

Leads to GS 38

5 a) iii) Describe, in detail, how the test would be carried out    (6 marks)

**Answer**
- Supply on
- Local isolator off
- Access live terminals in the local isolator
- At the incoming terminals of the local isolator test
  - L1 to isolator earthing terminal
  - L2 to isolator earthing terminal
  - L3 to isolator earthing terminal
- Close the local isolator cover
- Record highest result

**Comments**

The structure of this question is intended to help the candidate identify all relevant information.

A list has been used to answer a) iii) because it is easy to write down, easy to check and time efficient. Each statement must include sufficient information to make it clear how the test would be carried out. It is not necessary to include the instrument title and lead requirement as part of this answer as they have already been identified in i) and ii) above.

The test is carried out between each line conductor and the isolator earthing terminal. If a candidate described the isolator earthing terminal as "the earth terminal" "earth" or "the cpc" this would also gain the marks.

If the wrong test was described in the answer, such as $R_1 + R_2$ test and the result is then added to $Z_e$, then no marks would be awarded. The question specifically asks for an earth fault loop impedance test and not any test method that could be used to determine $Z_s$.

Answers that include dangerous procedures would score zero marks.

5 b) i) Determine, using the information in Figure 1 and 2, the expected $Z_s$ value for the circuit. Show all calculations.    (3 marks)

**Answer**

$$Z_s = Z_e + R_1 + R_2$$
$$= 0.11 + \frac{2 \times 7.41 \times 15}{1000}$$
$$= 0.11 + 0.22$$
$$= 0.33 \ \Omega$$

**Notes**

## Comments

Marks are awarded for each stage of the calculation. Setting out the calculation and showing each step makes it easy for the examiner to follow the process. It also helps the examiner award marks when the answer is not fully correct. Answers that are not clear, do not contain all steps and contain errors are likely to score fewer marks than an answer containing errors but is clearly laid out. This is because in the latter case the examiner can easily identify the correct parts of the calculation.

5  b)  ii)  Explain, in detail, why the measured value for this
            circuit is lower than the value calculated in b)i) above.    (2 marks)

## Answer

Resistance of $R_2$ is reduced due to steel conduit being in parallel with the cpc

## Comments

Although the answer is only worth 2 marks, the question does state "in detail". The examiner is looking for a complete answer. Answers such as "parallel paths" will not score 2 marks as this is not in detail.

6)  Describe, with the aid of a fully labelled diagram, the earth fault loop path for the outside lighting circuit. (15 marks)

## Answer

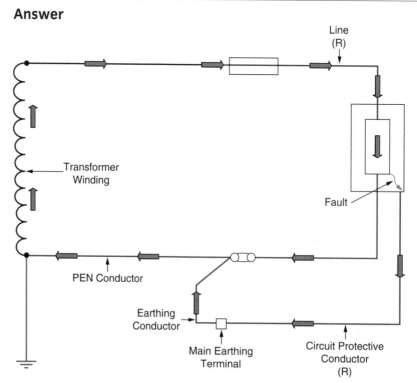

**TN-C-S System Earth Fault**

The fault path is:

From the point of fault
Along the cpc to the MET
From the MET via the earthing conductor to the supplier's PEN terminal
Along the PEN conductor to the supply transformer winding
Through the transformer winding, along the line conductor to the fault

**Comments**

This answer contains both a fully labelled diagram and a description of the path. Providing the diagram is very clear and the path is shown on the diagram, as is the case above, then full marks would be awarded. The description of the path gives additional information for the examiner to consider when awarding marks. It is in the interest of the candidate to give as much information as possible. If an incorrect system is described/drawn then no marks would be awarded. This is also true if the diagram is incomplete and the type of system cannot be determined.

Notes

## Written examination 2395-302

# Principles, practices and legislation for the periodic inspection of electrical installations (2395-302)

In this paper you will often be asked to provide longer answers for questions that ask you to 'describe' or 'explain'. These are often related to inspection and/or test procedures and you are required to demonstrate your knowledge of the inspection and/or test process.

For example:
*Describe, in detail, the procedure for carrying out a test to confirm the continuity of the main protective bonding conductor connected to the water installation pipework.* *(10 marks)*

There are a number of pointers in this question. The answer is worth 10 marks and the question asks for "detail", so a full description is required.

As this examination only considers periodic inspection, you can determine that the installation is energised. You would always need to confirm that it is safe to isolate the installation or circuit and state that safe isolation is carried out. On occasions the question may state that the permission to isolate has been given or safe isolation has been carried out.

Such questions may also be preceded by a number of auxiliary questions relating to the actions required before or after the test concerned. In such cases these items would not need to be repeated in the procedure.

There is no such indication for this question and so the requirements must be included and so you will need to include obtaining permission and the isolation procedure in your description.

### Option 1
The answer could be provided as a description of the process in a 'story' format, as in the following.

*"I would obtain permission from the client to isolate the installation, carry out safe isolation of the whole installation, lock off and retain the unique key. I would then disconnect one end of the main protective bonding conductor. I would select a low resistance ohmmeter, check condition and function, select a suitable long test lead and null the test leads. I would connect one test lead to the disconnected main protective bonding conductor and the other lead to the far end of the conductor. Test and record the result. I would then reconnect the main protective bonding conductor before reenergising the supply."*

**WRITTEN EXAMINATION 2395-302**
**PRINCIPLES, PRACTICES AND LEGISLATION FOR THE PERIODIC**
**INSPECTION OF ELECTRICAL INSTALLATIONS (2395-302)**

95

**Notes**

This answer and similar variations are perfectly acceptable and you would obtain the marks for the question. However, it does rely on the procedure being written down correctly the first time without missing any steps. It involves a considerable amount of writing and it is often difficult to spot any errors during the examination when you read back through your answer.

### Option 2

The answer can be structured as a set of bullet points, as follows.
* *Obtain permission to isolate the supply*
* *Safely isolate, lock off and retain the key*
* *Disconnect the main protective bonding conductor at one end*
* *Select a low resistance ohmmeter, check condition and function*
* *Select a suitable long test lead*
* *Null the test leads*
* *Connect the test leads to the disconnected bonding conductor and the far end*
* *Test and record the result*
* *Reconnect the main protective bonding conductor before reenergising the supply*

There is plenty of space to write down your answer, so you can leave a space between each bullet point, giving you the opportunity to add an additional line in later if you have missed a step in the process. You can also read through your answer much quicker and identify any omissions.

Additional bullet points may be added at the end if you find you have missed out some information providing they identify when the action is carried out. For example if you find you have omitted to null the test leads the final bullet point could be added as:

* *Before carrying out the test the test leads should be nulled*

This would then ensure that the appropriate marks are awarded for your answer.

The second option, with the use of the bullet points, is generally quicker to complete, less prone to errors or omissions and clearly demonstrates your understanding of the test process. You can try both methods to find out which best suits you. However, Option 2 is strongly recommended.

It is important to recognise the need to carry out testing in a way that does not put yourself or anyone else at risk. Also the test must be carried out in such a manner that produces valid results. No marks will be awarded for an answer that describes dangerous techniques and a heavy penalty will be applied to answers that describe invalid procedures.

The installations being inspected and tested are generally energised, unless information is given to the contrary. As a result the need to isolate in order to carry out certain tests must be included in the test process. The description would generally need to include: the need to obtain permission, to safely isolate and lock off and to retain the key. A description of the actual safe isolation process would only need to be provided where expressly asked for in the question.

This qualification relates to the periodic inspection of electrical installations and so encompasses the preparation, client discussion and the inspection, testing and reporting process. The testing process will include the need to confirm that test results are compliant with the requirements of BS 7671. Candidates will also need to identify non-compliances and award appropriate classification codes. Candidates will also be expected to identify the safety and practical aspects of periodic inspection.

### Inspection
The inspection process is a vital part of the periodic inspection process, many conditions may be identified during the inspection which would not be revealed by testing alone. Candidates are expected to be aware of the areas to be inspected, the actual items to be checked and the human senses to be used whilst inspecting those items. Recording the outcome for each inspection item is also a requirement of the reporting process.

Candidates should be aware that the Schedule of Inspections for the periodic inspection of installations provided in BS 7671 and IET GN 3, together with the examples of items requiring inspection given in BS 7671, Appendix 6, provides detailed information on the items of inspection for these installations. Candidates should become familiar with the items they are to consider, inspect and record and this will greatly improve both their understanding of the inspection process and their success in any related questions.

### Confirmation of compliance
Confirmation of compliance of the installation with the requirements of BS 7671 forms part of the periodic inspection process. The source document generally identifies that all testing is carried out at an ambient temperature of $20\,^{\circ}C$. This means that test values will not be at normal operating temperature and so this must be considered and where necessary compensated for.

# Sample test 2395-302

The sample exam paper below has six questions, the same number as the written examination, and follows the structure of the exam. The exam appears with spaces for you to write your answers so you can use it as a mock exam, if you run out of space continue on an additional sheet of paper. The exam is then repeated with worked-through answers and comments. The front cover of the examination paper is shown below, identifying the requirements of the exam. The date of the exam would normally appear on the front sheet and the heading "Exam Success" would be replaced with the examination series. The remaining information appears as it would on the examination paper.

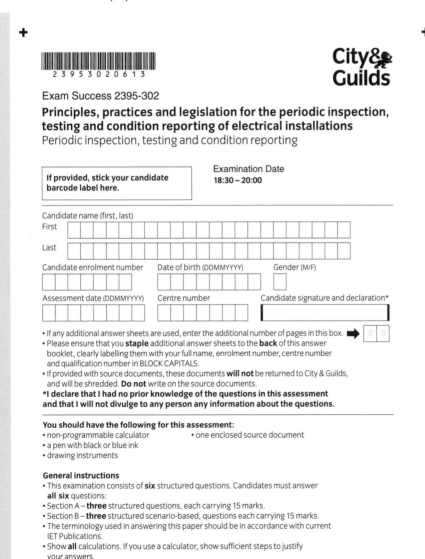

Exam Success 2395-302

**Principles, practices and legislation for the periodic inspection, testing and condition reporting of electrical installations**
Periodic inspection, testing and condition reporting

| If provided, stick your candidate barcode label here. | Examination Date 18:30 – 20:00 |

Candidate name (first, last)
First

Last

Candidate enrolment number     Date of birth (DDMMYYYY)     Gender (M/F)

Assessment date (DDMMYYYY)     Centre number     Candidate signature and declaration*

- If any additional answer sheets are used, enter the additional number of pages in this box.  ➡ 0 0
- Please ensure that you **staple** additional answer sheets to the **back** of this answer booklet, clearly labelling them with your full name, enrolment number, centre number and qualification number in BLOCK CAPITALS.
- If provided with source documents, these documents **will not** be returned to City & Guilds, and will be shredded. **Do not** write on the source documents.
**\*I declare that I had no prior knowledge of the questions in this assessment and that I will not divulge to any person any information about the questions.**

**You should have the following for this assessment:**
- non-programmable calculator          • one enclosed source document
- a pen with black or blue ink
- drawing instruments

**General instructions**
- This examination consists of **six** structured questions. Candidates must answer **all six** questions:
- Section A – **three** structured questions, each carrying 15 marks.
- Section B – **three** structured scenario-based, questions each carrying 15 marks.
- The terminology used in answering this paper should be in accordance with current IET Publications.
- Show **all** calculations. If you use a calculator, show sufficient steps to justify your answers.
- Write all your working out and answers in this booklet.

### General instruction
- This examination consists of **six** structured questions. Candidates must answer **all six** questions:
- Section A – **three** structured questions, each carrying 15 marks.
- Section B – **three** structured scenario-based questions, each carrying 15 marks.
- The terminology used in answering this paper should be in accordance with current IET publications.
- Show **all** calculations. If you use a calculator, show sufficient steps to justify your answers.

**Section A** – All questions carry equal marks. Answer **all three** questions. Show **all** calculations.

1   The electrical installation in a craft workshop is scheduled for a periodic inspection and test for insurance purposes.

   a)  State **two** statutory documents which apply to the inspection and testing process.    2 marks

_____

_____

_____

_____

   b)  State **three** non-statutory documents which the inspector may need to refer to relating specifically to inspection and testing.    3 marks

_____

_____

_____

_____

_____

_____

   c)  State who the inspector will consult to establish the extent and limitations of the periodic inspection and test.    2 marks

_____

_____

_____

_____

   d)  Explain why the sequence of testing for the periodic inspection may be different to that given in BS 7671 for initial verification.    4 marks

_____

_____

_____

_____

_____

_____

_____

_____

e) State the first action to be taken by the inspector should an exposed live part be identified on a socket-outlet circuit during the inspection.

4 marks

_____

_____

_____

_____

_____

_____

_____

2a A periodic inspection is to be undertaken in a hotel fitness room.
   i)   State **three** checks that need to be carried out to avoid damage to equipment before an insulation resistance test is conducted.

3 marks

_____

_____

_____

_____

_____

   ii)  State the test voltage to be applied for the insulation resistance test.

1 mark

_____

_____

   iii) State the effect on the value of insulation resistance produced by cable length.

1 mark

_____

_____

2b Insulation resistance tests between live conductors and earth on the individual circuits produced the following results.

$$200 \text{ M}\Omega, \; 200 \text{ M}\Omega, \; 150 \text{ M}\Omega, \; 50 \text{ M}\Omega, \; 25 \text{ M}\Omega, 100 \text{ M}\Omega \text{ and } 2 \text{ M}\Omega$$

i) Calculate the expected overall value of insulation resistance if the installation was tested with all the circuits connected. Show all calculations.      5 marks

ii) State, giving reasons, whether the insulation resistance for the installation complies with BS 7671.      3 marks

2c The lighting comprises a number of discharge luminaires.
State **two** alternative methods of preparing this circuit for the insulation resistance test.      2 marks

3a i) Explain the cause of voltage drop within an installation.                    3 marks

_____

_____

_____

_____

_____

_____

ii) State the two methods of determining voltage drop.                             2 marks

_____

_____

_____

_____

3b i) A radial circuit has a load current $I_b$ of 28 A at 230 V ac and has a combined
live conductor resistance of 0.16 Ω at 20 $^0$C. Determine the voltage drop for this circuit.
Show all calculations.                                                             5 marks

_____

_____

_____

_____

_____

_____

_____

_____

_____

ii) If the radial circuit supplies a machine lathe, determine whether the voltage drop in b i)
above complies with BS 7671.                                                       5 marks

_____

_____

_____

_____

_____

_____

_____

_____

_____

_____

**Section B –** All questions carry equal marks. Answer **all three** questions. Show **all** calculations.

**Questions 4 to 6 all refer to the enclosed scenario, see source document. Ensure you read this scenario before attempting these questions. Answers you provide must reflect the detail and information given in the scenario.**

4   a)   Describe **three** actions the inspector can take to ensure that no danger
occurs to those persons using the premises during the inspection and test.     3 marks

_____

_____

_____

_____

_____

_____

  b)   State the documents to be completed by the inspector and given to the client on
completion of the inspection and test.     3 marks

_____

_____

_____

_____

_____

_____

  c)   An inspection is being carried out within the distribution board.
    i)   State **three** areas relating to the circuit breakers that are to be inspected.     3 marks

_____

_____

_____

_____

_____

_____

_____

ii)  State what is being looked for by the inspector for each item in i) above.    3 marks

_____

_____

_____

_____

_____

iii) State the human sense that will be used during each check in ii) above .    3 marks

_____

_____

_____

_____

_____

5a Explain, giving reasons, whether a ring final circuit continuity test is required for Circuit 9.    3 marks

_____

_____

_____

_____

_____

5b A test of continuity of the ring final circuit supplying the stage area socket-outlets is to be carried out.

    i)    State the test instrument to be used for the test.     1 mark

_____

_____

    ii)   State **one** check to be made on the test instrument before carrying out the test other than calibration.     1 mark

_____

_____

    iii) State the action to be taken with the test leads before the test is carried out.     1 mark

_____

_____

5c Describe how the continuity of ring final circuit test is to be carried out on Circuit 9 once the circuit has been safely isolated and secured.     9 marks

_____

_____

_____

_____

_____

_____

_____

_____

_____

_____

_____

_____

_____

_____

_____

_____

_____

_____

_____

6a State whether the earthing conductor is either connected or disconnected when
carrying out tests to establish the prospective fault current.                    1 mark

_____

_____

6b Describe how a test is carried out to determine the prospective
fault current at the origin of the installation by direct measurement.            10 marks

_____

_____

_____

_____

_____

_____

_____

_____

_____

_____

_____

_____

_____

_____

_____

_____

_____

_____

_____

6c Explain how the inspector confirms that the installation complies with the requirements for prospective fault current protection.

4 marks

_____

_____

_____

_____

_____

_____

_____

## Source Document

Scenario (Section B – Questions 4 to 6)
(Source Document – Do not return to City and Guilds)
Do not write on here

The electrical installation in a 15 year old village hall requires inspection for a local authority license to hold functions. The building is used for various group activities each morning during the week and this is to continue during the inspection and test.

The supply and installation form part of a single-phase 230 V TN-S system having a $Z_e$ and PFC of 0.35 Ω and 1.0 kA respectively.

All circuits are installed using thermoplastic 70 °C single core cables, having copper conductors enclosed in surface mounted, PVC conduit and trunking. The cpcs are the same csa as the live conductors throughout.

Figure 1 shows information taken from the circuit schedule which is adjacent a metal-clad distribution board containing a mixture of Type B, BS EN 60898 circuit breakers and BS EN 61009 RCBOs.

One additional ring final circuit was installed approximately five years ago to supply socket-outlets in the stage area. There is no evidence of any other alterations or additions to the installation. The certification from the initial verification of the original installation and suitable circuit charts are available for the inspector. There is no certification available for the additional ring final circuit

Metallic oil and water installation pipework is installed within the building and 10 mm² main protective bonding conductors are installed within the building fabric and connected to the pipework.

All testing is carried out at a temperature of 20 °C.

Figure 2 on next page shows the distribution layout .

## Figure 1

| Circuit No. | Device rating | Description | Conductor csa in mm² | |
|---|---|---|---|---|
| | | | Live | cpc |
| 1 | 32 A | Ring final circuit for socket-outlets main hall | 2.5 | 2.5 |
| 2 | 32 A | Ring final circuit for socket-outlets other areas | 2.5 | 2.5 |
| 3 | 32 A | Cooker | 6.0 | 6.0 |
| 4 | 16 A | Immersion heater | 2.5 | 2.5 |
| 5 | 16 A | Boiler | 2.5 | 2.5 |
| 6 | 10 A | Lights main hall | 1.5 | 1.5 |
| 7 | 10 A | Lights other areas | 1.5 | 1.5 |
| 8 | 6 A | Outside lights | 1.5 | 1.5 |
| 9 | 32 A* | Ring final circuit stage area | 2.5 | 2.5 |
| 10 | - | Spare | - | - |

* indicates BS EN 61009 type B RCBO

## Figure 2: Distribution layout

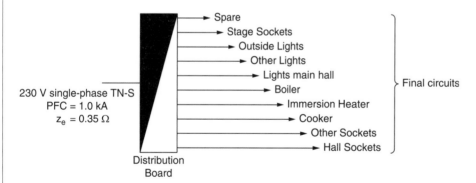

# Questions and answers

The questions in Sample test 2395-302 are repeated below with sample answers, and comments and advice where appropriate.

1) The electrical installation in a craft workshop is scheduled for a periodic inspection and test for insurance purposes.

   a) State **two** statutory documents which apply to the inspection and testing process. (2 marks)

   **Answer**
   - Electricity at Work Regulations
   - Health and Safety at Work (etc) Act

   **Comments**
   Abbreviations such as EWR and HSWA are acceptable. Candidates must correctly identify whether these are an Act or Regulations in order to achieve the marks.

1 b) State **three** non-statutory documents which the inspector may need to refer to relating specifically to inspection and testing. (3 marks)

   **Answer**
   - BS 7671
   - On-Site Guide
   - Guidance Note 3

   **Comments**
   GS 38 is an acceptable alternative so any three from these four are acceptable. Abbreviations such as OSG and GN 3 are also acceptable answers. As the question asks for three items, the examiner will only mark the first three responses.

1 c) State who the inspector will consult to establish the extent and limitations of the periodic inspection and test. (2 marks)

   **Answer**
   - The client
   - The insurance company

   **Comments**
   The person ordering the work **or** the person requesting the work are acceptable alternatives to the client. Note: Read the question carefully, the main question stem specifies that the work is for insurance purposes.

**Notes**

The response must relate to the insurance company and so other interested third parties or any other specific third party such as licensing authority etc. will not receive a mark.

1   d)   Explain why the sequence of testing for the periodic inspection may be different to that given in BS 7671 for initial verification.   (4 marks)

### Answer

The installation has been inspected and tested and placed in service. Periodic inspection verifies the current condition or the safety of the installation and not the confirmation of the safety of the installation before placing in service.

### Comments

The candidate needs to explain the purpose of periodic inspection and how this differs from initial verification in order to obtain the marks. Examiners will consider the description used and whether this explains the reason for there being no sequence for the periodic tests. Answers such as inconvenience and because the installation is energised are not acceptable as these do not affect the sequence of testing but relate to the extent and limitations agreed with the client.

Tip: Many candidates confuse the sequence of tests with the sampling of the installation. Sampling does not relate to the sequence, BS 7671 gives a required sequence for the tests to be carried out at initial verification. All these tests could be carried out at periodic inspection but not necessarily in the same sequence. The answer needs to clarify why this is the case.

1   e)   State the first action to be taken by the inspector should an exposed live part be identified on a socket-outlet circuit during the inspection.   (4 marks)

### Answer

Make a recommendation to the responsible person that the circuit is isolated until a repair can be carried out.

### Comments

Alternative answers include;

Inform client and obtain permission to isolate until a repair can be carried out

Or

Isolate with the client's permission until a repair can be carried out.

'Isolate the supply and record the results' and similar answers are not acceptable as the inspector is required to inform and recommend the isolation. Not to carry out the isolation without first seeking agreement from the client/user.

Guidance Note 3, in the general requirements section, identifies the action to be 'recommend the immediate isolation of the defective part.' The person ordering the work should be informed, in writing, of this recommendation without delay. Answers reflecting this requirement would also be accepted.

2 a) A periodic inspection is to be undertaken in a hotel fitness room.
   i) State **three** checks that need to be carried out to avoid damage to equipment before an insulation resistance test is conducted. (3 marks)

**Answer**
Link out sensitive equipment, disconnect loads, disconnect pilot lamps/indicators

**Comments**
Alternative answers include; ensure no connection between live conductors and earth **or** link out controls.

As the question asks for three items, the examiner will only mark the first three responses.

2 a) ii) State the test voltage to be applied for the insulation resistance test (1 mark)

**Answer**
500 volts

2 a) iii) State the effect on the value of insulation resistance produced by cable length (1 mark)

**Answer**
As length increases resistance decreases

**Comments**
An alternative answer includes as length decreases resistance increases. Alternative terms such as higher or lower are acceptable.

2  b)  Insulation resistance tests between live conductors and earth on the individual circuits produced the following results. 200 MΩ, 200 MΩ, 150 MΩ, 50 MΩ, 25 MΩ,100 MΩ and 2 MΩ.

    i)  Calculate the expected overall value of insulation resistance if the installation was tested with all the circuits connected. Show all calculations. (5 marks)

**Answer**

$$\frac{1}{R_T} = \frac{1}{R_1} + \frac{1}{R_2} + \frac{1}{R_3} + \frac{1}{R_4} + \frac{1}{R_5} + \frac{1}{R_6} + \frac{1}{R_7}$$

$$\frac{1}{R_T} = \frac{1}{200} + \frac{1}{200} + \frac{1}{150} + \frac{1}{50} + \frac{1}{25} + \frac{1}{100} + \frac{1}{2}$$

$$\frac{1}{R_T} = 0.005 + 0.005 + 0.007 + 0.02 + 0.04 + 0.01 + 0.5$$

$$\frac{1}{R_T} = 0.587$$

$$R_T = 1.7 \text{ M}\Omega$$

**Comments**

Either the formula or the figures are acceptable for the first line. The correct units must be included in the final answer to obtain the marks.

Tip: It is important to show all the calculations used to arrive at the answer as should an error be made the marks lost will relate just to the error whereas providing just the answer or omitting some calculation stages together with obtaining the wrong answer will result in all the marks being lost.

2  b)  ii)  State, giving reasons, whether the installation resistance for the installation complies with BS 7671. (3 marks)

**Answer**

1.7 MΩ ≥ than minimum 1 MΩ and so it complies.

**Comments**

Candidates need to explain the reason for their answer so simply 'yes' would only receive 1 mark.

The response to Part ii) of this question is dependent on the answer to the candidates response to Part i). If the candidate has an incorrect answer in i) this will not affect the marks awarded in ii). The examiner will mark ii) based upon the answer given by the candidate in i).

**Notes**

2 c) The lighting comprises a number of discharge luminaires.
State two alternative methods of preparing this circuit for the
insulation resistance test. (2 marks)

**Answer**
Disconnect all luminaires or link live conductors and then test live
conductors to Earth.

**Comment**
An alternative answer for one of these is to switch off at the local switch.

Tip: When testing insulation resistance the measurements must be taken
to Earth and disconnecting and testing to a cpc will result in marks being
lost.

3 a) i) Explain the cause of voltage drop within an
installation. (3 marks)

**Answer**
Volt drop is a product of the conductor resistance and the load current.

**Comment**
The question refers to the volt drop that occurs normally within the
installation which is caused by the conductor resistance and the current
flowing. Do not confuse this with the causes of excess voltage drop due to
poor design or overloading of the circuit.

3 a) ii) State the two methods of determining voltage drop. (2 marks)

**Answer**
Measurement and Calculation.

**Comments**
It is not acceptable to carry out a direct measurement of voltage drop
using volt meter(s). The circuit conductors must be at their normal
operating temperature, the circuit under full load and there must be no
variation in the supply voltage during the test. The methods given are as
a result of the measurement of conductor resistance and the reference
to charts or tables giving the details of voltage drop. These are not the
figures in the tables in Appendix 4 of BS 7671, which are generic design
details for the calculation of appropriate cable sizes.

**Notes**

3  b)  i)  A radial circuit has a load current $I_b$ of 28 A at 230 V ac and has a combined live conductor resistance of 0.16 Ω at 20 °C. Determine the voltage drop for this circuit. Show all calculations.  (5 marks)

### Answer

Voltage drop $= (R_1 + R_n) \times Ib \times 1.2$

so Voltage drop $= 0.16 \times 28 \times 1.2 = 5.376$ V

### Comments

The use of the conductor resistance and load current together with the 1.2 multiplier to compensate for the difference in conductor temperature at the time of test and the normal operating temperature of the conductors when maximum resistance and hence maximum volt drop will occur.

3  b)  ii)  If the radial circuit supplies a machine lathe, determine whether the voltage drop in b i) above complies with BS 7671.  (5 marks)

### Answer

Maximum volt drop = 230 V x 5% = 11.5 V.

As 5.376 V is equal to or less than 11.5 V so complies.

### Comments

Alternative calculations such as:

$$\text{max volt drop} = 5\% \text{ Vdrop} = \frac{5.376}{230} \times 100 = 2.3\% \text{ which is less than 5\%}$$

so OK,

may be used and will attract the same marks.

Tip: It is important to show the calculations because in this type of question where an error is made in the calculation candidates are only penalised once for the error. The remainder of the answer is marked based upon the incorrect figure produced and, providing the process is correct, marked accordingly. Candidates could have completely the wrong answer for 3 b i) but based upon their incorrect figure from that calculation achieve full marks for 3 b ii) providing their calculation and conclusion is correct.

6  b)   Describe how a test is carried out to determine the
         prospective fault current at the origin of the installation
         by direct measurement.                                    (10 marks)

**Notes**

### Answer
- Secure the area around distribution board.
- Access incoming live terminals.
- Using a PFC tester (or EFLI tester set to kA).
- Confirm test leads comply with GS 38.
- Connect to incoming supply side Line and Earth.
- Measure PEFC.
- Connect to incoming supply side Line and Neutral.
- Measure PSCC .
- Record highest result as the PFC .
- Reinstate the DB.

### Comments
In this case there have been no previous questions relating to the
preparation such as type of instrument etc. and so this information needs
to be included in the answer.

Tip: Failure to identify that the test leads must comply with GS 38 will lose
several marks. It is vital to remember that the earthing and all protective
bonding conductors are connected whilst this test is carried out.

6  c)   Explain how the inspector confirms that the installation complies
         with the requirements for prospective fault current protection.   (4 marks)

### Answer
Confirm measured PFC is no greater than the breaking capacity (or $I_{cn}$) of
the protective devices.

### Comments
An alternative response would be: Protective device $I_{cn}$ is equal to or more
than the measured PFC.

Notes

# More information

More information

**Notes**

# Further reading

Regulatory requirements, Standards and publications are constantly being updated. It is important to ensure that any reference material you use to support your knowledge and understanding is the latest edition and is still current.

*IET Wiring Regulations 17th Edition (BS 7671:2008 incorporating amendment number 1:2011)*, published by the IET, London

*On-Site Guide (BS 7671:2008 Wiring Regulations, incorporating amendment number 1:2011)*, published by the IET, London

IEE Guidance Notes, a series of guidance notes, each of which enlarges upon and amplifies the particular requirements of a part of the *IET Wiring Regulations, Seventeenth Edition*, published by the IET, London:
– Guidance Note 1: *Selection and Erection of Equipment*
– Guidance Note 2: *Isolation and Switching*
– Guidance Note 3: *Inspection and Testing*
– Guidance Note 4: *Protection Against Fire*
– Guidance Note 5: *Protection Against Electric Shock*
– Guidance Note 6: *Protection Against Overcurrent*
– Guidance Note 7: *Special Locations*
– Guidance Note 8: *Earthing and Bonding*

**Notes**

**Notes**